Michael Shoichedbrod
Irina Shoichedbrod

24-hour operation of bomb shelters

Michael Shoichedbrod
Irina Shoichedbrod

24-hour operation of bomb shelters

Using solar and fiber optics to provide 24-hour operation bomb shelter and electric vehicle traffic

ScienciaScripts

Imprint

Cover image: www.ingimage.com

This book is a translation from the original published under ISBN 978-620-8-43813-5.

Publisher:
Sciencia Scripts
is a trademark of
Dodo Books Indian Ocean Ltd. and OmniScriptum S.R.L publishing group

120 High Road, East Finchley, London, N2 9ED, United Kingdom
Str. Armeneasca 28/1, office 1, Chisinau MD-2012, Republic of Moldova, Europe
Managing Directors: Ieva Konstantinova, Victoria Ursu
info@omniscriptum.com

Printed at: see last page
ISBN: 978-620-8-84953-5

Contents

Dedicated

To our family:

Elena and Eden, Igor, Julia and Sarochka, Arik and Reshelle

INTRODUCTION

The use of solar energy and fibre optics to provide round-the-clock lighting for bomb shelters, their oxygen, hydrogen and electricity life-support systems

A sunflower shaped device, a light receiver that dynamically tracks the light activity of sunlight collected throughout the day, allows the collected sunlight to be channelled via fibre optics into bomb shelters from where it is distributed to different rooms [1].

The internal room lighting fixtures transmit sunlight with all its morning, afternoon and evening variations in colour and intensity, and are duplicated by light-emitting diodes (LEDs) that are powered at night by electricity from photocell batteries charged during the day from the lighting fixtures.

Previously, the author developed a 24/7 continuous hydrogen and oxygen generation in a specially designed photoelectrolyser-generator, achieved by using a lamp as the illumination for the photoelectrolyser-generator, and an LED with a battery charged during lamp operation as a useful electrical load for the photoelectrolyser-generator, which starts illuminating the photoelectrolyser-generator when the lamp is switched off until the battery is completely discharged, and continuously filling the photoelectrolyser-generator

Use of solar energy and optical fibre to ensure uninterrupted movement of the electric vehicle

The round-the-clock production of oxygen and hydrogen in a specially designed photoelectrolyser-generator enabled the author to perform round-the-clock and continuous charging of the fuel cell of an electric vehicle for its uninterrupted driving [2].

In this book, the author has revealed the mechanism of 24-hour solar lighting of bomb shelters, using a light receiver that dynamically tracks the light activity of sunlight collected during the day, transmitted via optical fibre to the bomb shelters to illuminate the rooms, which charges batteries of photovoltaic cells located in the most illuminated areas of these rooms, the electric current from which at night turns on LEDs located near the interior lights of the rooms, until the batteries are fully discharged in the morning.

The 24-hour, continuous production of hydrogen and oxygen of the bomb shelter life support system in a specially designed photoelectrolyser-generator is achieved by using as light for the photoelectrolyser-generator sunlight from a light receiver transmitted via optical fibre, and as a useful electrical load for the photoelectrolyser-generator a battery-operated LED which is charged during the day by illuminating the photoelectrolyser-generator with sunlight which starts to illuminate the photoelectrolyser-generator.

Round-the-clock production of oxygen and hydrogen in a specially designed photoelectrolysator-generator enables 24-hour, continuous charging of the fuel cell of the life support system's electrical generator for uninterrupted power supply to bomb shelters.

Round-the-clock, continuous production of hydrogen and oxygen in a specially designed photoelectrolyser-generator, achieved by using as lighting of the photoelectrolyser-generator sunlight from a light-receiver transmitted via optical fibre, and LED with a battery, charged during the day by illumination of the photoelectrolyser-generator by sunlight, which begins to illuminate the photoelectrolyser-generator in the absence of sunlight at night until the battery is completely discharged in the morning, and not

CHAPTER I

The use of solar energy and fibre optics to

The use of solar energy and fibre optics to provide round-the-clock lighting for bomb shelters, their

life-support systems with oxygen, hydrogen,

electricity

1.1 Use of solar energy and fibre optics to provide round-the-clock illumination of bomb shelters

The sunflower type solar receiver is a device consisting of 36 Fresnel lenses that rotate evenly around their axis inside the device, following the sun throughout the day (Figure 1, [1]).

Figure 1: A system for natural lighting of any building using sunlight passing through fibre optics, developed by the Swedish company Parans in close cooperation with scientists at the University of Technology.

Dynamic tracking of light activity is achieved by an integrated photosensor, microprocessor and motors, whose total power consumption does not exceed 10W. The light receiver is assembled for 6000 lumens.

One device is enough to illuminate a room of 30-40 m^2, the external unit weighs 30 kg and is installed above the surface of the Earth.

The indoor lighting units let in sunlight with all its morning, afternoon and evening variations of colour and power, but the invisible spectrum, including infrared and ultraviolet radiation, is filtered in such a way that both the fading of things and the possibility of human sunburn are excluded.

The sunlight collected during the day is transported to the bomb shelters via fibre optics, where it is distributed to the different rooms.

An optical fibre is a thin round glass filament designed to transmit light [3-6]. Structurally, an optical fibre is a coaxial structure consisting of a core and a frame coated with an acrylate compound to protect it from the external environment. An optical fibre consists of a core (called mandrel in some literature) and a sheath made of fused silica (Figure 2).

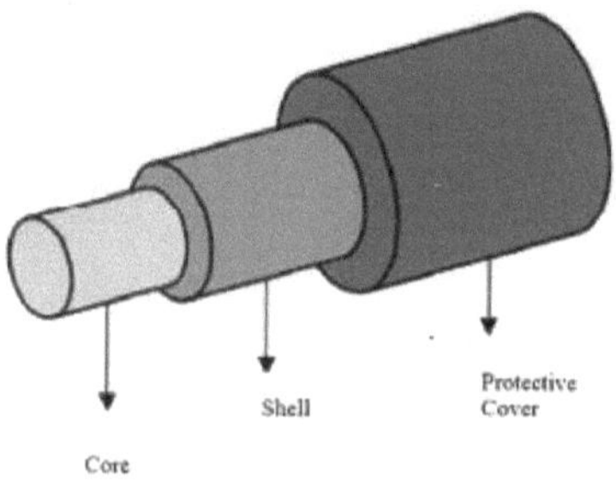

Figure 2: Optical fibre

Although both parts of the fibre are made of fused silica glass, they have different refractive indices: the core has a higher refractive index and the cladding has a lower refractive index.

The core of the fibre is used to transmit the light flux, while the cladding is needed to create conditions for the phenomenon of total internal reflection. Internal reflection is the phenomenon of reflection of an electromagnetic wave (light is also an electromagnetic wave) from the interface between two media (marked in Figure 3).

Due to total internal reflection, the light flux inside an optical fibre is able to travel considerable distances - from hundreds of metres to hundreds of kilometres.

Optical fibres are divided into two groups: multimode and singlemode (Figure 4).

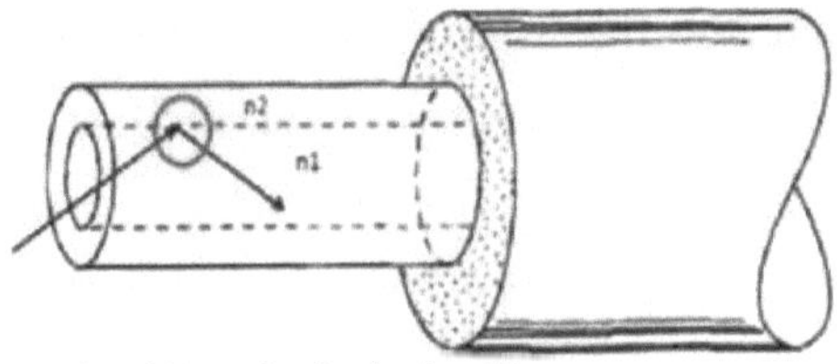

Figure 3: The phenomenon of total internal reflection in a fibre

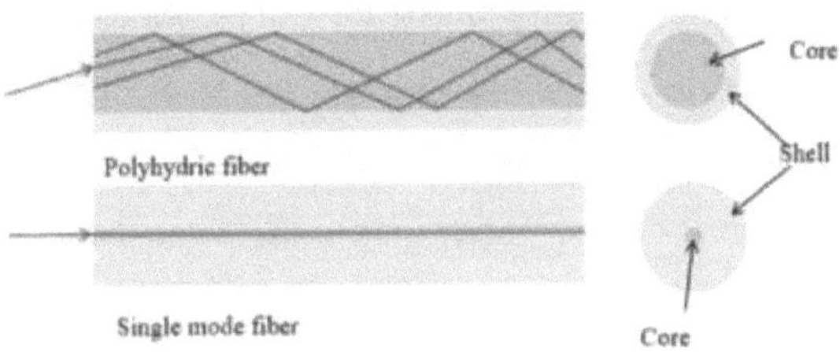

Figure 4: Multimode and single-mode optical fibres

Structurally, they differ only in the diameter of the core: in multimode fibres it is larger than the wavelength of the transmitted light signals, so that several optical modes (rays) are transmitted along the fibre, while in single-mode fibres the core is smaller than the wavelength of the light signal and this allows only one mode to be transmitted.

The scope of natural light propagation through fibre optics is wider than that of solar panels, limited by low height, trajectory and the presence of internal free space for the pipe, which is bulkier than thin and inconspicuous fibre optic cables.

In addition, fibre optic solar lighting can be switched on and off with a simple switch that allows the lenses to be turned away from the sunlight.

Sunlight through fibre optics creates better illumination, allows for more efficient use of darkened rooms, and has been proven to improve people's well-being, normalise their biological clocks and increase productivity. In addition,

20 per cent of all electricity consumed worldwide is spent on artificial lighting, including during daylight hours.

Thanks to a fibre optic solar lighting system, the use of artificial lighting can be halved, which means that CO2 emissions can be reduced regionally and internationally and global warming can be combated.

The system under development implements round-the-clock solar illumination of bomb shelters by means of a light receiver dynamically tracking the light activity of sunlight collected during the day, transmitted via fibre optics to underground structures directly to illuminate the rooms, which charges batteries of photovoltaic cells located in the most illuminated areas of these rooms, the electric current from which at night switches on LEDs located next to the internal lighting fixtures of the rooms until the batteries are completely discharged. The interior lighting fixtures let in sunlight with all its morning, afternoon and evening variations in colour and power, but the invisible spectrum, including infrared and ultraviolet radiation, is filtered in such a way that both the fading of things and the possibility of a person getting a tan are eliminated.

Indoor lighting of premises is carried out with the help of fibre optic luminaires [7- 9].

Fibre optic luminaire consists of: optical cable, light source (sunlight) and terminals - light guides.

The terminals are crystals, lenses that allow to direct the light from the optical cable.

Figure 5 shows examples of fibre optic lights.

Sunlight collected during the day is transmitted to the bomb shelters via fibre optics, where it is distributed to different rooms, using a fibre optic splitter, also called a beam splitter, which is an integrated waveguide optical distribution device capable of splitting an incident light beam into two or more light beams (Figure 6).

Room lighting charges batteries of photocells located in the most illuminated areas of these rooms, the electric current from which at night switches on the LEDs located next to the indoor room lighting until the batteries are fully discharged in the morning.

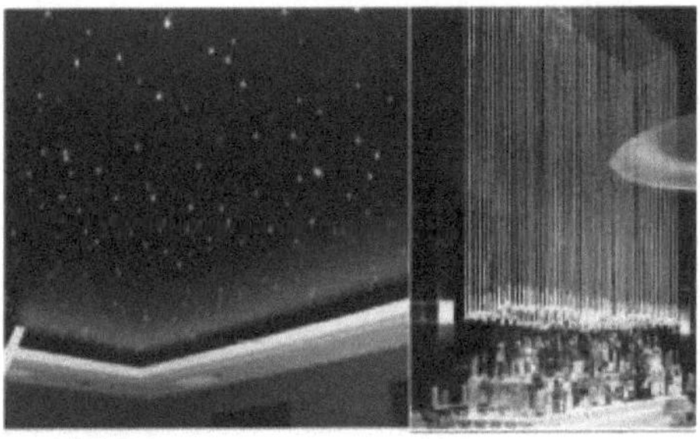

Figure 5: Examples of fibre optic lamps

Figure 6: A splitter, which is an integrated waveguide optical distribution device capable of splitting an incident light beam into two beams.

A photocell is a device that produces photocurrent under the influence of the sun [10-12].

A semiconductor photocell has a stable structure.

The atoms are firmly bonded by covalent bonding.

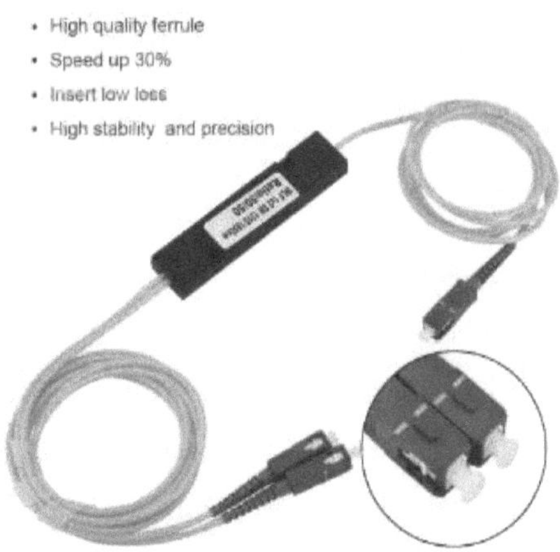

Figure 6: A splitter, which is an integrated waveguide optical distribution device capable of splitting an incident light beam into two beams.

If the energy of the quantum is sufficient to break the bond between the electron and the atom, the electron becomes free.

In its place, a hole (a positive charge equal to the electron charge) is born.

If a potential difference is applied, an electric current is generated.

The weak conductivity of pure semiconductors is increased by adding other substances to produce more free positively or negatively charged particles:

1. impurities that create an excess of electrons form an n-type semiconductor.
2. impurities creating an excess of holes will form a p-type semiconductor.

A double layer of charges, called a P-N junction, and an electric field that prevents their further redistribution will arise (Figure 7).

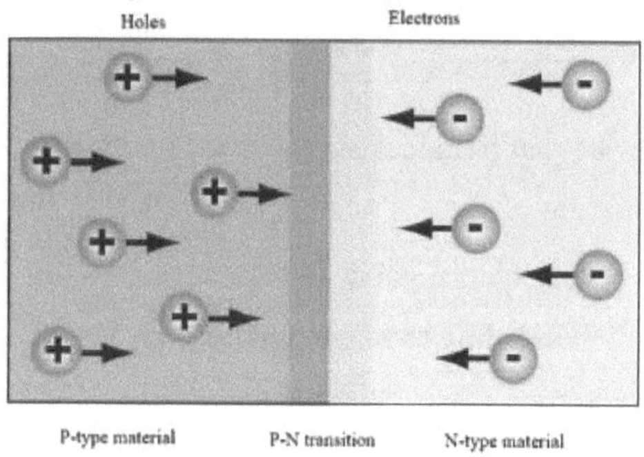

Figure 7: Creation of P-type material, P-N junction and N-type material in a semiconductor photocell

If the circuit is closed, an electric current will be induced that is proportional to:

1. The intensity of the light flux;
2. the area of the semiconductor;
3. duration of luminescence.

The structure of a semiconductor photocell is as shown in Figure 8.

As a useful electrical load of the photocell is used LED with a battery, charged during the day by illuminating the photocell with sunlight collected during the day and transmitted via optical fibre to the bomb shelters directly to illuminate the premises, the electric current from which at night turns on the LEDs located near the interior lighting of the premises until the batteries are fully discharged in the morning.

Typically, the photocell is combined with an LED with a battery charged during the day by illuminating the photocell with sunlight, in the form of a table lamp, which includes (Figure 9):

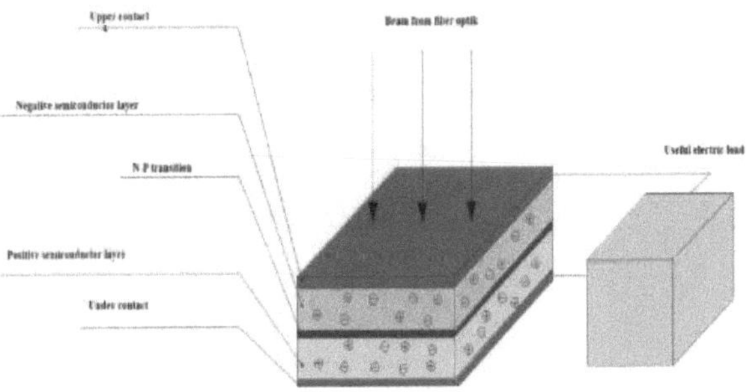

Figure 8: Structure of the semiconductor photocell

- safety glasses;
- solar cell (several photocells);
- photoresistor;
- charging battery;
- control board (for controlling switch on of battery charging and switch off LED at day light and switch off of battery charging and switch on LEDby night);
- LED;
- lamp body

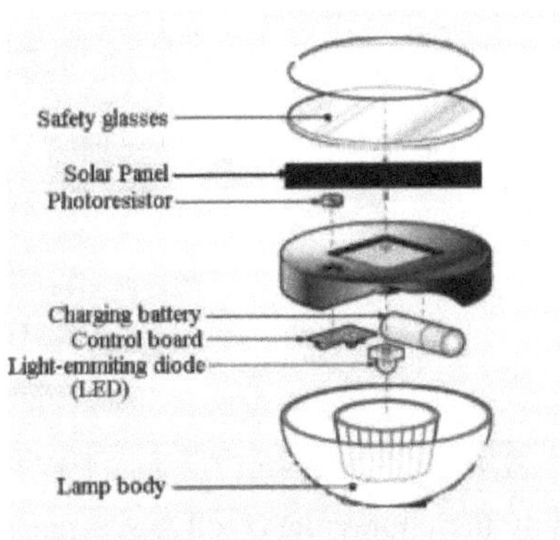

Figure 9: Solar powered LED table lamp for indoor use

Figure 10 shows a commercial version of the LED solar-powered indoor table lamp.

Figure 11 shows a commercial version of a solar-powered LED ceiling lamp for indoor use in which the solar panel (photovoltaic cells) is connected separately to the LED ceiling lamp, which corresponds to our case.

Figure 10: Commercial version of the solar-powered indoor LED table lamp

Figure 11: Commercial version of an indoor LED solar ceiling lamp with a solar panel. luminaire in which the solar panel (photovoltaic cells) is connected separately to the LED ceiling luminaire

The circuit diagram of the solar LED control board is shown in Figure 12.

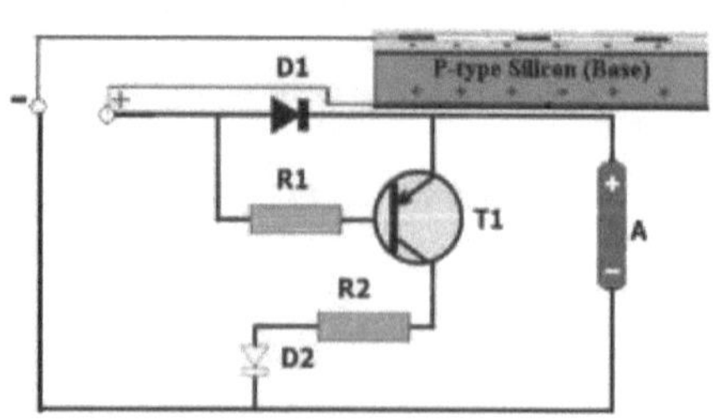

Figure 12: Circuit diagram of the control board of the solar LED downlight.

The current generated by the photovoltaic cell (solar panel) charges the battery (A) through diode D1.

The positive potential applied to the base through resistor R1 "holds" transistor T1 in the off state and LED D2 does not glow.

When the illumination of the solar panel decreases significantly, the transistor opens (due to a decrease in the positive potential applied to the base) and connects LED D2 to the battery.

The LED starts to glow, illuminating the rooms of the bomb shelter.

Diode D1 prevents the battery from discharging during light illumination.

When the light illumination is switched on, the positive voltage coming from the "+" terminal of the solar panel to the base "closes" the transistor T1 and the LED D2 stops to light up and the battery starts to charge.

Figure 13 schematically shows the transmission of sunlight collected during the day to the shelter via fibre optics, where it is distributed to the different rooms using a fibre optic splitter.

Figure 13: The sunlight collected during the day is transmitted to the bomb shelters via fibre optics, where it is distributed to different rooms using a fibre optic splitter.

1.2 The use of solar energy and fibre optics to to provide a 24-hour life support system of bomb shelters with oxygen, hydrogen.

In the developed system, round-the-clock, continuous production of hydrogen and oxygen for the underground life-support system is carried out in the previously developed photoelectrolyser-generator by using in this case as light for the photoelectrolyser-generator sunlight from the light-receiver transmitted via optical fibre, and as a useful electrical load of the photoelectrolyser-generator - LED with a battery, charged in the process of illumination of the photoelectrolyser-generator with sunlight for a period of time.

A completely new type of photoelectrolyser-generator developed earlier by the author is capable of producing separately pure hydrogen and oxygen at below market price.

In the developed photoelectrolyser-generator, which has a combined anode, which is a layer of N-type silicon semiconductor with a large-cell metal grid attached to it, immersed in ordinary water, under the influence of four photons of light (sunlight from a light receiver transmitted via optical fibre) four electrons are "knocked out" leaving four "holes" which, when released, "attract" four electrons from four hydroxyl anions $-4OH^-$ and two molecules of dissociated water:

$$4OH^- - 4e = 2H_2O + O2\uparrow, \qquad (1.2.1)$$

forming two water molecules and an oxygen molecule, which float off the mesh anode as a bubble of gaseous oxygen and are discharged from the photoelectrolyser through a separate oxygen outlet tube.

The remaining two positively charged hydrogen cations (H^+) of the dissociated water, due to the negatively charged cathode, move towards it and, on reaching it, return two electrons to it, neutralising into two separate hydrogen atoms which, being in a free state, combine with each other to form a gaseous hydrogen moleculc:

$$2H^{+} + 2e^{-} \rightarrow H_2 \uparrow, \qquad (1.2.2)$$

which in water takes the form of an electrolytic bubble of gaseous hydrogen, which floats up from the cathode as a bubble of gaseous hydrogen and is discharged from the photoelectrolyser generator through a separate hydrogen discharge tube.

A direct electric current is generated for the electrons "knocked out" from the pooled anode of the silicon layer of the N-type semiconductor with the attached large-cell metal grid and two electrons returned to the cathode from the two hydrogen cations, directed from the combined anode of the N-type silicon semiconductor layer with the attached coarse mesh metal grid to the cathode, thereby generating a potential difference between the electron-rich cathode and the electron-depleted combined anode of the N-type silicon semiconductor layer with the attached coarse mesh metal grid.

As a result of this potential difference, the photoelectrolysis process of ordinary water occurs, resulting in the formation and removal of hydrogen gas bubbles from the cathode and photoelectrolyser-generator through a separate hydrogen removal tube, and the formation and removal of oxygen gas bubbles from the combined anode of the silicon layer N-type semiconductor layer with a coarse-mesh metal grid and photoelectrolyser-generator through a separate oxygen removal tube (Figure 14).

Location of the electrolysis base (1), the main element of the developed photoelectrolyser-generator, including electrodes with a thin membrane (2) located between them and a special mechanism for regulating the gap between them, at the bottom of the photoelectrolyser-generator.

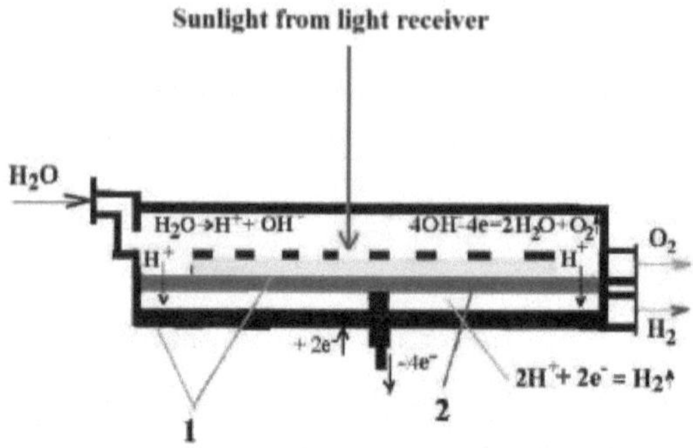

***Figure** 14: Continuous operation of the developed photoelectrolyser that generates and discharges hydrogen and oxygen bubbles separately through nozzles.*

The main features of the developed photoelectrolyser-generator, in contrast to the known photoelectrochemical cells [13, 14], are: horizontally located anode made of silicon semiconductor with attached metal mesh with large cells and annealed graphite cathode (electrolysis base) in the lower part of the photoelectrolyser-generator; a membrane made of fire hose material located between the electrodes; a device regulating the gap between the electrodes.

A thin membrane made of fire hose material placed between a combined anode made of silicon semiconductor with an attached metal mesh with large cells. and the cathode in the electrolysis base prevents the penetration of electrolytic oxygen bubbles formed as a result of the attraction of four hydroxyl anions of the surrounding dissociated water, giving up four electrons to four positively charged "holes", formed by the action of four photons of light on the combined anode made of silicon semiconductor.

As a result of electrolysis in the cathode zone of the photoelectrolyser, electrolytic hydrogen bubbles are intensively formed, which are removed as a foam of gaseous hydrogen from the cathode zone of the photoelectrolyser through a hydrogen tube.

The arrangement of the electrolysis base, i.e. electrodes with a thin membrane and a small gap between them, in the lower part of the photoelectrolyser leads, in contrast to the photoelectrochemical cells existing today, where the electrodes are arranged vertically, to an intensive release of hydrogen and oxygen bubbles without the use of expensive electrolytes.

The arrangement of electrodes in vertical direction, as it happens in photoelectrochemical cells with aqueous electrolyte solution, in contrast to the electrolysis base located in the bottom of the photoelectrolyser, where, as in all electroflotation devices, the electrodes are arranged in horizontal direction, occupying the entire bottom of the electroflotation apparatus, sharply reduces the cross-sectional area and increases the length of the conductive layer between the electrodes.

The efficiency of a photoelectrochemical cell or photoelectrolyser is determined by the intensity of gas release in them.

The gas release (M) at the electrode during the time (t), according to Faraday's law, is calculated by the following formula:

$$M = jigSt \qquad (1.2.3)$$

where j - electrochemical equivalent of gas (for hydrogen j= 1,0-10-7 kg/(A^s); for oxygen j– 8,29-10-8 kg/(A^s)); i - current density, A/m^2; g= 90%, 95% - current yield (ratio of actual gas mass transfer to theoretical calculated one in per cent); S - cross-sectional area of conducting layer between electrodes, m^2.

A sharp decrease in the cross-sectional area and a sharp increase in the length of the conductive layer between the electrodes in accordance with (1.2.3) leads to a sharp decrease in the outgassing of hydrogen and oxygen in the photoelectrochemical cell and, as a consequence, to a sharp decrease in the efficiency of the photoelectrochemical cell.

To compensate for the reduction of hydrogen and oxygen outgassing in the aqueous electrolyte solution, expensive electrolyte (e.g., ***NaOH***) is used, the concentration of which in the solution is 40%, which leads to a sharp increase in energy consumption of the photoelectrochemical cell, reaching 4.5 kW^h/m3 for hydrogen production.

A significant advantage of the developed photoelectrolyser-generator over existing electroflotation devices is the absence of electrodes in the electrolysis base, as which are used dense meshes with fine cells, causing high currents of about 100 amperes at low voltages (20-30 V).

The explanation of this significant advantage is as follows.

If we consider the ***i*** - process current density as ***I/S***, where ***I*** is the direct process current, equation (1.2.3) takes the following form:

$$M = jIgt \qquad (1.2.4)$$

From equation (1.2.4) it can be seen that the mass of gas released at the electrode during time t is directly proportional to the magnitude of the direct current and is independent of the area of the electrode.

However, according to Ohm's law, the magnitude of the DC current of the process is directly proportional to the electrode voltage (***U***) and inversely proportional to the resistance of the interelectrode medium (***R***):

$$I = U/R \qquad (1.2.5)$$

The resistance of the interelectrode medium (***R***), in turn, depends on the interelectrode gap (***L***) and the cross-sectional area of the conductive layer between the electrodes (***S***) as follows::

$$R = \rho\frac{L}{S} \qquad (1.2.6)$$

Where ρ is the resistivity of the inter-electrode medium.

Substituting (1.2.6) into (1.2.5), we obtain:

$$I = (US/\rho L) \qquad (1.2.7)$$

Thus, it can be seen from (1.2.7) that a decrease in the cross-sectional area of the conductive layer between the electrodes of the photoelectrochemical cell (***S***), in our case a change in the shape of the electrode (silicon semiconductor with attached mesh anode), can be compensated by increasing the voltage (***U***) at the electrodes to maintain the same efficiency of the photoelectrochemical cell (the same gas content of the system - **M**).

Increasing the voltage at the electrodes not only does not reduce the efficiency of the photoelectrolyser-generator, but also, as it was said earlier, provides maximum outgassing of intensively formed hydrogen bubbles at the cathode and oxygen at the combined anode made of silicon semiconductor and metal mesh with large cells.

The arrangement of electrodes in the vertical direction not only sharply reduces the cross-sectional area and the length of the conductive layer between the electrodes, which leads to a sharp decrease in gas emission and intensity of formation of electrolytic bubbles of hydrogen and oxygen, but also prevents the formation and rapid removal of electrolytic bubbles of hydrogen and oxygen through the hydrogen and oxygen tubes, respectively.

Electrolytic bubbles of hydrogen or oxygen formed on the entire horizontal surface of the electrode or anode, breaking off from the electrode surface, rush upwards and in the form of gaseous hydrogen bubbles or oxygen bubbles are removed from the photoelectrolyser-generator through hydrogen or oxygen tubes due to Archimedes' force acting in a strictly vertical direction with respect to the cathode or anode.

In vertically arranged electrodes, the hydrogen and oxygen bubbles formed float near the cathode and anode, forming a narrow flow due to the Archimedes force acting tangentially to the cathode or anode, preventing the hydrogen bubbles from breaking away from the cathode or the oxygen bubbles from the anode.

In the developed photoelectrolyser-generator the 24-hour continuous production of hydrogen and oxygen for the underground life support system is carried out by the same technology as described above, i.e. by using in this case as lighting of the photoelectrolyser-generator the sunlight from the light receiver transmitted via optical fibre, and as a useful electrical load of the photoelectrolyser-generator - LED (3) with a battery, charged in the process of lighting of the photoelectrolyser-generator by the sunlight.

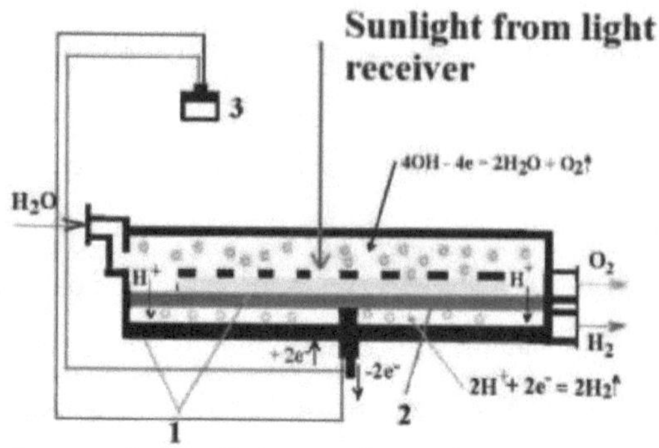

Figure 15: 24/7 continuous production of hydrogen and oxygen for an underground life support system

In this case, an LED with a battery charged in the process of illuminating the photoelectrolyser-generator with sunlight during the day is used as the useful electrical load of the photoelectrolyser-generator, which starts to illuminate the photoelectrolyser-generator in the absence of sunlight at night until the battery is completely discharged in the morning.

In this case, the circuit diagram of the LED control board will look as shown in Figure 16.

The current produced by the photoelectrolyser-generator charges the battery (A) through diode D1.

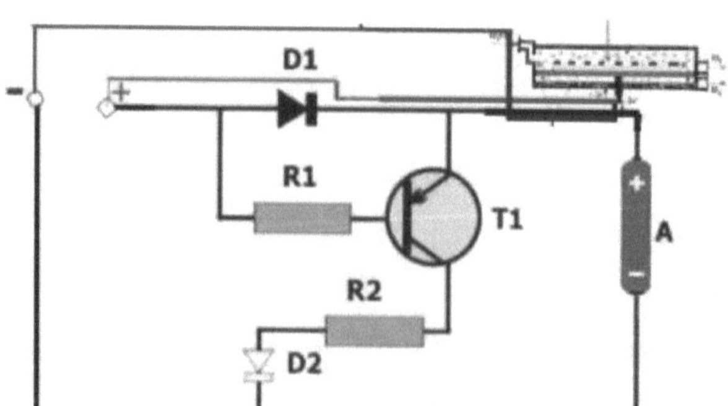

Figure 16: Circuit diagram of the LED light control board.

The positive potential applied to the base through resistor R1 "holds" transistor T1 in the off state and LED D2 does not glow.

When the illumination of the anode of the photoelectrolyser-generator anode decreases significantly, the transistor opens (due to a decrease in the positive potential applied to the base) and connects the LED D2 to the battery.

The LED starts to glow, illuminating the anode of the photoelectrolyser-generator.

Diode D1 prevents the battery from discharging during light illumination.

When the light illumination is switched on, the positive voltage coming from the "+" pin of the anode of the photoelectrolyser-generator anode to the base "closes" transistor T1 and LED D2 stops glowing and the battery starts to charge.

Figure 17 schematically shows the 24-hour, continuous hydrogen and oxygen generation for an underground life support system in the previously developed photoelectrolyser-generator using sunlight from a light receiver transmitted via optical fibre as light for the photoelectrolyser-generator.

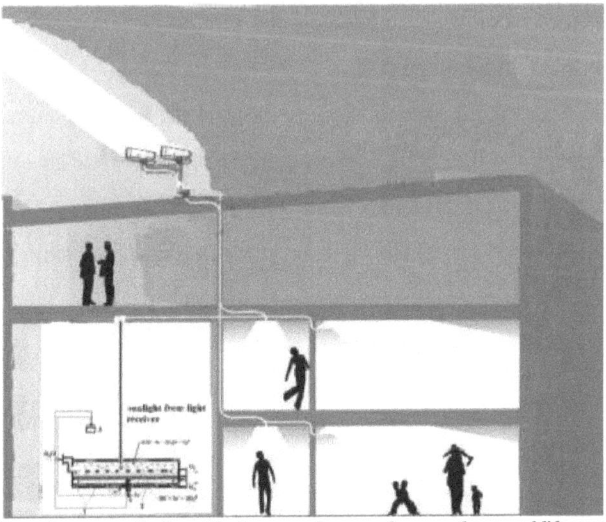

Figure 17: 24/7, continuous production of hydrogen and oxygen for an underground life support system in a previously developed photoelectrolyser-generator using sunlight from a light-receiver transmitted via optical fibre as light for the photoelectrolyser-generator.

1.3 The use of solar energy and fibre optics to provide 24-hour power supply for life support systems.

bomb shelters

The developed round-the-clock and continuous production of hydrogen and oxygen in a photoelectrolyser-generator enabled the author to provide round-the-clock and continuous charging of the fuel cell of the engine of the electrical generator of the bomb shelter life support system for its uninterrupted operation, provided, as previously described, by using sunlight from a light receiver transmitted via optical fibre as the lighting of the photoelectrolyser-generator, and light as the electrical payload of the photoelectrolyser-generator

An electric generator is a device that works by converting mechanical energy into electrical energy. It works on the principle of electromagnetic induction, which is to create an electric current by moving a wire next to a magnet.

However, in order for induction to produce a practical amount of electricity, you need a strong and constant force that moves the wire fast enough. This is where mechanical energy comes into play.

Figure 18 shows how the mechanical motion of a magnet into and out of a coil creates a magnetic field that counteracts the motion, which creates an induced alternating current and voltage.

In our case, the hydrogen and oxygen produced by the photoelectrolyser-generator charges the fuel cells of the engine of the electrical generator, which provides the mechanical force needed to generate the current.

The engine rotates a shaft, which rotates an electromagnet (armature).

The rotating armature rotates within a stationary magnetic field (stator) to generate electric current through copper wiring.

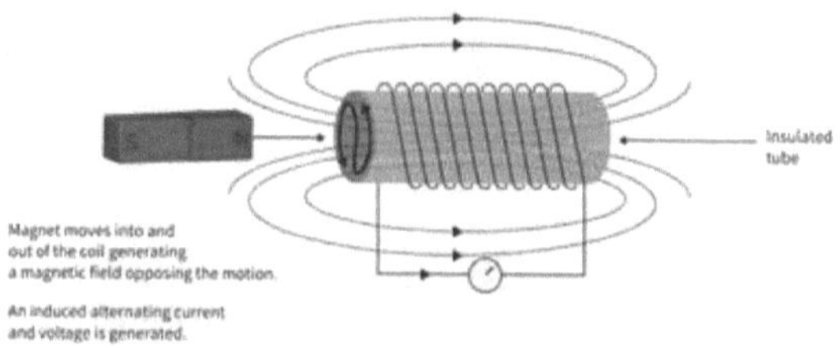

Figure 18: The mechanical movement of the magnet in and out of the coil, creating a magnetic field opposite to the motion that produces the induced alternating current and voltage.

Essentially, a motor drives a rotating shaft that turns an armature and generates electricity.

The motor rotates the shaft, which in turn rotates the electromagnet (armature).

The rotating armature rotates within the stationary magnetic field (stator) to generate electric current through the copper wiring.

That is, the motor drives a rotating shaft that rotates the armature and generates electricity.

Figure 19 shows a general view of an electric generator.

Thus, the process of using solar energy and fibre optics for 24-hour power supply of bomb shelters is divided into two stages: 24-hour and continuous production of hydrogen and oxygen in the photoelectrolyser generator and 24-hour and continuous charging of the fuel cell of the engine of the electric generator of the bomb shelter life support system for its uninterrupted operation and 24-hour power supply of the bomb shelter.

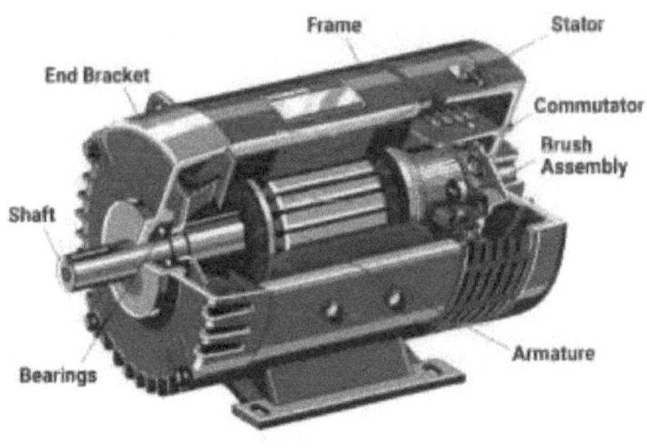

Figure 19: General view of the electrical generator

Using the mechanism theoretically described by the author [2] for the round-the-clock process of photoelectrolysis of water under the influence of sunlight energy on a combined silicon semiconductor anode with attached metal grid of photoelectrolyser-generator, a new photoelectrolyser-generator powered by sunlight energy was developed to produce pure hydrogen and oxygen round-the-clock, which charges the fuel cell of the engine of the electric generator for its round-the-clock operation.

The hydrogen and oxygen in the developed generator are produced by electrolysis of ordinary water continuously fed into the photoelectrolyser-generator, at below market cost, due to a specially designed electrolysis base including a fire membrane placed between the electrodes and a mechanism for adjusting the gap between the electrodes located at the bottom of the photoelectrolyser-generator.

Round-the-clock production of oxygen and hydrogen in the photoelectrolyser-generator, and consequently round-the-clock continuous charging of the fuel cell of the engine of the electric generator for its uninterrupted operation is provided by using as illumination of the photoelectrolyser-generator sunlight from a light-receiver transmitted via optical fibre, and as a useful electrical load of the photoelectrolyser-generator - a light-emitting diode with an accumulator charged in the process of illumination of the photoelectrolyser-generator with sunlight.

The photoelectrochemical cell is a silicon semiconductor with an attached mesh anode and a cathode, connected by an electrical circuit for flowing a direct electric current generated when light is applied to a silicon semiconductor with an attached mesh anode immersed in an aqueous electrolyte solution. The phosphorus-added N-type silicon layer has a shared outer orbital crowded with electrons, resulting in the tendency of the phosphorus-added N-type silicon layer to give up electrons to bring the shared orbital to a normal state, so the combined silicon semiconductor and metal mesh anode, immersed in an aqueous electrolyte solution, excited by four photons of light, "loses four electrons, leaving four positively charged "holes" that attract four electrons from the four hydroxyl anions ($4OH^-$) of the neighbouring dissociated into a hydrogen cation (H^+) and a hydroxyl anion (OH^-) of water:

$$H_2O \leftrightarrow H^+ + OH^-, \quad (1.3.1)$$

resulting in the formation of two water molecules and an oxygen molecule, which in the aqueous electrolyte solution takes the form of an oxygen electrolysis bubble floating from the cathode to the free surface of the aqueous electrolyte solution:

$4OH^- - 4e^- = 2H_2O + O_2\uparrow$ (1.3.2)

Two positively charged hydrogen cations $2H^+$ of the neighbouring dissociated aqueous solution reaching the negatively charged cathode, the negative charge of the cathode is due to the freed four electrons flowing from the combined silicon semiconductor anode with attached metal grid through the load to the cathode, returning two electrons to it, neutralising into two separate hydrogen atoms in the free state, combining with each other to form a gaseous hydrogen molecule which, in an aqueous electrolyte solution, takes the form of an electrolytic hydrogen bubble floating from the cathode to the free surface of the aqueous electrolyte solution.

$2H^+ + 2e^- \rightarrow H_2\uparrow$ (1.3.3)

Four electrons "knocked out" from the combined silicon semiconductor anode and metal grid, passing through the N type layer of the combined silicon semiconductor anode with the metal grid attached thereto, along an electrical circuit to the cathode, and two electrons, returned to the cathode from two hydrogen cations make it the negative pole and the anode losing electrons becomes the positive pole, generating a direct electric current from the combined silicon semiconductor anode and metal grid to the cathode (Figure 20).

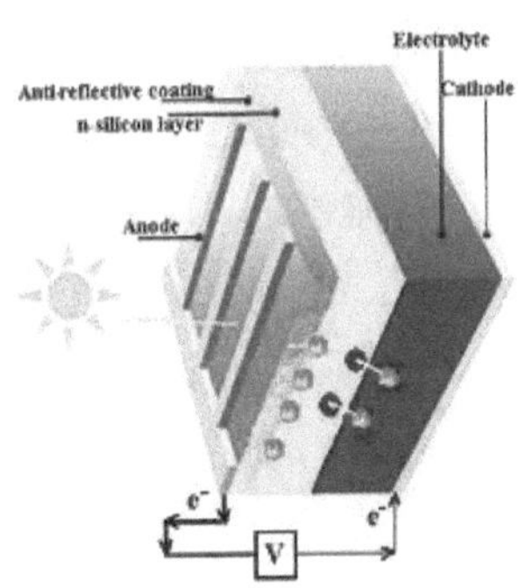

Figure 20: The combined N-type silicon semiconductor anode with attached metal grid operates under sunlight.

As established earlier [2], the size and intensity of the formation of negatively charged electrolytic hydrogen bubbles and positively charged oxygen bubbles can be controlled by increasing the electrical potential of the electrodes, making the intensely formed bubbles as small as required by the process.

The conclusion stated above was the basis for the development of a new photoelectrolyser-generator.

An important feature of the developed photoelectrolyser-generator, which produces pure hydrogen and oxygen at a below-market price, is a specially designed electrolysis base, which includes a membrane of fire-retardant material located between a combined silicon semiconductor anode with an attached metal grid, and an annealed graphite cathode, and a mechanism for adjusting the electrode gap at the bottom of the photoelectrolyser-generator.

A thin fire hose membrane placed between the combined silicon semiconductor anode with attached metal mesh and the cathode in the electrolysis base prevents the penetration of electrolytic oxygen bubbles formed by the attraction of four hydroxyl anions of the surrounding dissociated water giving up four electrons into four positively charged "holes" formed when four photons of light are applied to the combined silicon semiconductor anode with attached metal mesh.

As a result of electrolysis in the cathode zone of the photoelectrolyser-generator, electrolytic hydrogen bubbles are intensively formed, which are removed from the cathode zone of the photoelectrolyser through a hydrogen tube in the form of a foam of gaseous hydrogen.

The arrangement of the electrolysis base, i.e. electrodes with a thin membrane and a mechanism for adjusting the gap between them, in the lower part of the photoelectrolyser-generator leads, in contrast to the photoelectrochemical cells existing today, where the electrodes are arranged vertically, to an intensive release of hydrogen and oxygen bubbles without the use of expensive electrolytes.

Arranging the electrodes vertically, as is the case in the

photoelectrochemical cells with aqueous electrolyte solution, unlike the electrolysis base located at the bottom of the photoelectrolyser-generator, where, as in all electroflotation devices, the electrodes are arranged horizontally, occupying the entire bottom of the electroflotation apparatus, sharply reduces the area of cross-section and increases the length of the conductive layer between the electrodes.

The efficiency of a photoelectrochemical cell or photoelectrolyser-generator is determined by the intensity of gas release in them.According to Faraday's law, the mass of gas (**M**) released at the electrode during time (*t*) is equal to:

$$M = jigSt \qquad (1.3.4)$$

where ***j*** - electrochemical equivalent of gas (for hydrogen j= 1,0-10-7 kg/(A-s); for oxygen ***j***= 8,29-10-8 kg/(A-s)); i - current density, A/m2; ***g*** = 90%, 95% - current yield (ratio of actual gas mass transfer to theoretical calculated one in per cent); ***S*** - cross-sectional area of conductive layer between electrodes, m2.

A sharp decreasc in the cross-sectional area and a sharp increase in the length of the conductive layer between the electrodes in accordance with (1.3.4) leads to a sharp decrease in the outgassing of hydrogen and oxygen in the photoelectrochemical cell or photoelectrolyser-generator and, as a consequence, to a sharp decrease in the efficiency of the photoelectrochemical cell or photoelectrolyser-generator.

To compensate for the reduction of hydrogen and oxygen outgassing in the aqueous electrolyte solution, expensive electrolyte (e.g., ***NaOH***) is used, the concentration of which in the solution is 40%, which leads to a sharp increase in energy consumption of the photoelectrochemical cell, reaching 4.5 kW^h/m3 for hydrogen and oxygen production.

A significant advantage of the developed photoelectrolyser-generator over existing electroflotation devices is the absence of electrodes in the electrolysis base, as which are used dense grids with fine cells, causing high currents of the order of 100 amperes at low voltages (20-30 V).

The explanation of this significant advantage is as follows.

If we consider the i - process current density as ***I/S***, where **I** is the constant process current, then equation (1.3.4) takes the following form:

$$M = jIgt \qquad (1.3.5)$$

From equation (1.3.5) it can be seen that the mass of gas released at the electrode during time t is directly proportional to the magnitude of the direct current and does not depend on the area of the electrode. However, according to Ohm's law, the magnitude of the direct current of the process is directly proportional to the electrode voltage (***U***) and inversely proportional to the resistance of the interelectrode medium (***R***):

$$I = U/R \qquad (1.3.6)$$

The resistance of the interelectrode medium (***R***), in turn, depends on the interelectrode gap (***L***) and the cross-sectional area of the conductive layer between the electrodes (***S***) as follows:

$$R = \rho\frac{L}{S} \qquad (1.3.7)$$

Where ρ is the resistivity of the inter-electrode medium.
Substituting (1.3.7) into (1.3.6), we obtain:

$$I = (US/\rho L) \qquad (1.3.8)$$

Thus, it is clear from (1.3.8) that a decrease in the cross-sectional area of the conductive layer between the electrodes of a photoelectrochemical cell or photoelectrolyser-generator (***S***), in our case a change in the shape of the electrode (silicon semiconductor with attached mesh anode), can be compensated by increasing the voltage (***U***) at the electrodes to maintain the same efficiency of the photoelectrochemical cell or photoelectrolyser-generator (previous gas content of the system - ***M***).
Increasing the voltage at the electrodes not only does not reduce the energy consumption efficiency of the photoelectrochemical cell, reaching 4.5 kW^h/m3 (for sunlight using photoelectrolyser-generator, the energy consumption efficiency reaches 0) to produce hydrogen and oxygen.
A significant advantage of the developed photoelectrolyser-generator over existing electroflotation devices is the absence of electrodes in the electrolysis

base, as which dense meshes with fine cells are used, causing large currents of the order of 100 amperes at low voltages (20-30 V).

The arrangement of electrodes in the vertical direction not only sharply reduces the cross-sectional area and the length of the conductive layer between the electrodes, which leads to a sharp decrease in the outgassing and intensity of the formation of electrolytic bubbles of hydrogen and oxygen, but also prevents the formation and rapid removal of electrolytic bubbles of hydrogen and oxygen through the hydrogen and oxygen tubes, respectively.

Electrolytic bubbles of hydrogen or oxygen formed on the entire horizontal surface of the electrode or anode, breaking away from the electrode surface, rush upwards and are removed as hydrogen gas bubbles or oxygen bubbles from the photoelectrolyser-generator through hydrogen or oxygen tubes due to the Archimedes force acting in a strictly vertical direction with respect to the cathode or anode.

In the vertically arranged electrodes, the formed hydrogen and oxygen bubbles float near the cathode and anode, forming a narrow flow due to the Archimedes force acting tangentially to the cathode or anode, preventing the hydrogen bubbles from breaking away from the cathode or the oxygen bubbles from the anode.

In the developed photoelectrolysator-generator is carried out round-the-clock, continuous production of hydrogen and oxygen for underground life support system, by the same technology as described above, i.e. by using in this case as lighting photoelectrolysator-generator sunlight from the light receiver, transmitted via optical fibre, and as a useful electrical load photoelectrolysator-generator - LED (3) with a battery, charged in the process of lighting photoelectrolysator-generator.

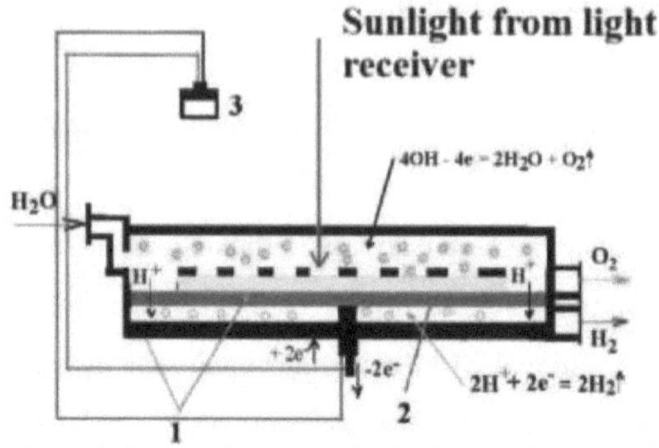

Figure 21: 24/7, continuous hydrogen and oxygen production for an underground life support system

In this case, the photoelectrolyser-generator electrical payload is an LED with a battery charged while illuminating the photoelectrolyser-generator with sunlight during the day, which starts illuminating the photoelectrolyser-generator in the absence of sunlight at night until the battery is fully discharged in the morning.

In this case, the circuit diagram of the LED control board will look as shown in Figure 22.

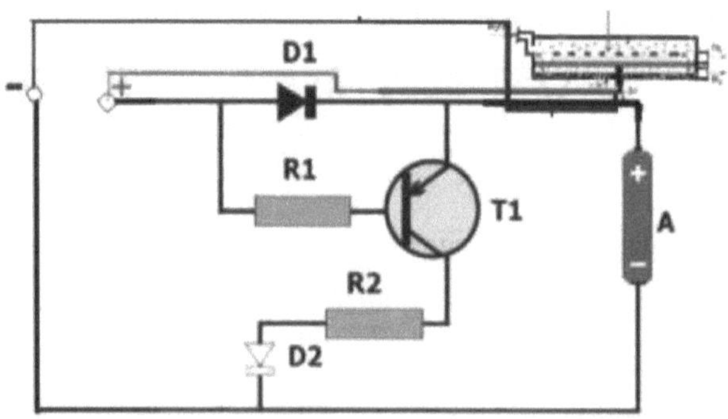

Figure 22: Circuit diagram of the LED control board.

The current produced by the photoelectrolyser generator charges the battery (A) through diode D1.

The positive potential applied to the base through resistor R1 "holds" transistor T1 in the off state and LED D2 does not glow.

When the illumination of the anode of the photoelectrolyser-generator anode decreases significantly, the transistor opens (due to a decrease in the positive potential applied to the base) and connects the LED D2 to the battery.

The LED starts to glow, illuminating the anode of the photoelectrolyser-generator.

Diode D1 prevents the battery from discharging during light illumination.

When the light illumination is switched on, the positive voltage coming from the "+" terminal of the photoelectrolyser-generator anode to the base "closes" the transistor T1 and the LED D2 stops glowing and the battery starts charging.

Continuous charging of the fuel cell of the engine of the electric generator motor of the bomb shelter life support system for its uninterrupted operation and round-the-clock supply of electricity to the bomb shelter is carried out as follows.

The resulting gases are separately fed hydrogen into the porous hydrogen electrode (anode) and oxygen into the porous oxygen electrode (cathode) of the fuel cell (Figure 23). The anode and cathode are separated by a membrane.

Each of the electrodes is coated with a layer of catalyst.

As a result, the four hydrogen molecules in the hydrogen electrode lose four electrons, creating a negative potential at one of the engine contacts, and, reacting with the four ***OH***anions obtained from the oxygen electrode by reacting the oxygen molecule with two water molecules, releasing four positively charged ones

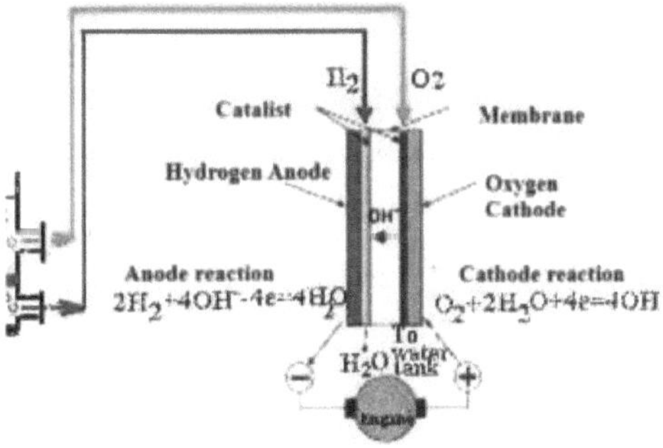

Figure 23: The resulting gases separately enter hydrogen into the porous hydrogen electrode (anode), oxygen into the porous oxygen electrode (cathode) of the fuel cell.

The particles, creating a positive potential at another contact on the engine, form four water molecules (water vapour):

$$O_2 + 2H_2O + 4e \rightarrow 4OH^- \rightarrow 2H_2 + 4OH^- - 4e \rightarrow 4H_2O \qquad (1.3.9)$$

As a result, the engine starts and the generated water vapour flows back to the conventional water tank, closing the cycle.

Thus, the developed 24/7 hydrogen generator produces pure hydrogen and oxygen, which continuously charges the (energy efficient) fuel cell of the engine's electric generator.

Figure 24 shows a complete cycle of 24/7 hydrogen and oxygen production and charging of the fuel cell of the electric engine generator fuel cell of the bomb shelter life support system using the previously developed photoelectrolyser-generator, which uses sunlight (3) from a light receiver transmitted via optical fibre as illumination of the photoelectrolyser-generator and an LED (4) with a battery charged during the illumination of the photoelectrolyser-generator with sunlight during the day, which starts o

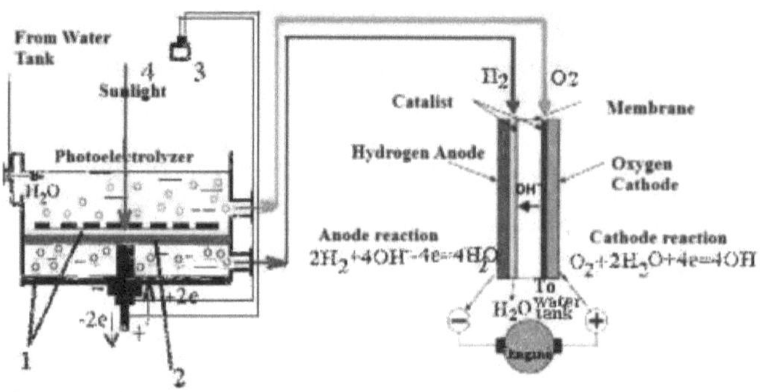

Figure 24: Complete cycle of 24/7 hydrogen and oxygen production and charging of the fuel cell of the bomb shelter life-support system's electric generator fuel cell

continuous charging of the fuel cell of the electric generator with the obtained oxygen and hydrogen; return of the water vapour released during charging of the fuel cell of the electric generator to the tank with ordinary water.

Thus, the developed system allows to provide round-the-clock illumination of all rooms of the underground structure, round-the-clock replenishment of the atmosphere of the bomb shelter with oxygen, as well as round-the-clock uninterrupted operation of the engine of the electric generator to ensure constant power supply of the bomb shelter.

Figure 25 shows a schematic representation of the operation of the electrical generator engine to provide continuous electrical power to the bomb shelter.

Figure 25 Operation of the electric generator engine to provide continuous power to the bomb shelter

CHAPTER II

The use of solar energy and optical fibre to provide a Uninterrupted movement of electric vehicle

The developed round-the-clock and continuous hydrogen and oxygen production in the photoelectrolyser-generator enabled the author to provide round-the-clock and continuous charging of the fuel cell of the vehicle's electric motor for uninterrupted movement of the vehicle, provided, as described earlier, by the use of sunlight from a light-receiver transmitted via optical fibre as illumination of the photoelectrolyser-generator from a light-receiver transmitted via optical fibre, and an LED as the useful electrical load of the photoelectrolyser-generator

A vehicle electric motor is a motor whose operating principle is based on the conversion of electrical energy into mechanical rotational energy.

The general view of the electric motor is shown in Figure 26.

The main components of a transport electric motor involved in energy conversion are the stator and rotor.

The principle of operation of any transport electric motor is based on the processes of mutual attraction and repulsion of the poles of magnets on the rotor and stator.

The movement is carried out under the action of the magnetic field itself and inertia.

The rotor is the rotating part of a transport electric motor consisting of a core, winding and shaft.

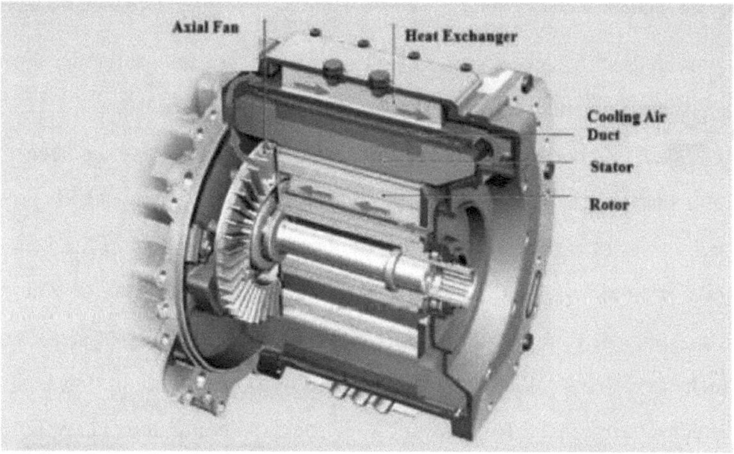

Figure 26:General view of an electric motor

The stator (inductor) is stationary. It is responsible for producing a stationary magnetic field.

The stator consists of a frame, core and winding:

The core is a metal rod with a winding on its periphery.

The magnetic circuit of the motor is closed directly through the core.

The core is made of round steel plates, similar to a layer cake.

Insulated steel plates with silicon additives are used in the manufacture of the cores, which provides an increase in the efficiency of the electric motor of the vehicle, the smallest specific metal losses per unit mass, and a reduction in the magnitude of demagnetising Foucault eddy currents arising from the remagnetisation of the core.

There are longitudinal grooves on the core surface. The winding is laid through them.

The shaft is a metal rod directly transmitting the torque, also made of electrical steel and serving as a base for fixing the core.

The ends of the shaft have threads, recesses for gears, rolling bearings, pulleys.

Stator frame - stator housing. As a rule, the frame is aluminium or cast iron.

The frame serves as a base for fixing the main and auxiliary poles.

Stator core - a cylinder made of profiled steel sheets. It is screwed inside the frame.

It is provided with slots for the winding, which creates magnetic flux and, crossing the rotor conductors, induces electromotive force in them.

In our case, the hydrogen and oxygen produced by the photoelectrolyser-generator charges the fuel cells of the vehicle's electric motor, which provides the mechanical force necessary for the uninterrupted movement of the vehicle.

Thus, the process of using solar energy and fibre optics to provide round-the-clock uninterrupted movement of the vehicle is divided into two steps: round-the-clock and continuous production of hydrogen and oxygen in the photoelectrolyser-generator and round-the-clock and continuous charging of the fuel cell of the vehicle motor for uninterrupted movement of the vehicle.

Using theoretically described by the author [2] mechanism of round-the-clock process of photoelectrolysis of water under the influence of sunlight energy on a combined anode made of silicon semiconductor with attached metal grid of photoelectrolyser-generator allowed to develop a new photoelectrolyser-generator operating on the energy of sunlight for round-the-clock production of pure hydrogen and oxygen, which charges the fuel cell of the electric generator engine for its round-the-clock operation.

Hydrogen and oxygen in the developed generator are produced by electrolysis of ordinary water continuously fed into the photoelectrolyser-generator, at a cost below the market cost, due to a specially designed electrolysis base, including a fire membrane placed between the electrodes and a mechanism for adjusting the gap between the electrodes, located in the lower part of the photoelectrolyser-generator.

Round-the-clock production of oxygen and hydrogen in the photoelectrolyser-generator, and consequently round-the-clock continuous charging of the fuel cell of the engine of the electric generator for its uninterrupted operation is provided by using as light for the photoelectrolyser-generator solar light from a light-receiver transmitted via optical fibre, and as a useful electrical load of the photoelectrolyser-generator - a light-emitting diode with an accumulator charged in the process of illumination of the photoelectrolyser-generator with solar light.

The photoelectrochemical cell is a silicon semiconductor with an attached reticulated anode and a cathode, connected by an electrical circuit for flowing a direct electric current generated when light is applied to a silicon semiconductor with an attached reticulated anode immersed in an aqueous electrolyte solution.

The N-type silicon layer augmented with phosphorus has a shared outer orbital crowded with electrons, resulting in the tendency of the N-type silicon layer with added phosphorus to give up electrons to bring the shared orbital to a normal state, so a combined silicon semiconductor and metal grid anode immersed in an aqueous electrolyte solution, excited by four photons of light, "loses four electrons, leaving four positively charged 'holes' that attract four electrons from the four hydroxyl anions (4OH-) of the neighbouring dissociated into a hydrogen cation (H^+) and a hydroxyl anion (OH^-) of water:

$$H_2O \leftrightarrow H^+ + OH^-, \qquad (2.1)$$

resulting in the formation of two water molecules and an oxygen molecule, which in the aqueous electrolyte solution takes the form of an oxygen electrolysis bubble floating from the cathode to the free surface of the aqueous electrolyte solution:

$$4OH^- - 4e^- = 2H_2O + O_2\uparrow \qquad (2.2)$$

Two positively charged hydrogen cations $2H^+$ of the neighbouring dissociated aqueous solution reaching the negatively charged cathode, the negative charge of the cathode is due to the freed four electrons flowing from the combined silicon semiconductor anode with attached metal grid through the load to the cathode, returning two electrons to it, neutralising into two separate hydrogen atoms in the free state, combining with each other to form a gaseous hydrogen molecule which, in an aqueous electrolyte solution, takes the form of a hydrogen electrolysis bubble floating from the cathode to the free surface of the aqueous electrolyte solution.

$2H^{+} + 2e^{-} \rightarrow H_2\uparrow$ (2.3)

Four electrons "knocked out" from the combined silicon semiconductor anode and metal grid, passing through the N-type layer of the combined silicon semiconductor anode with the metal grid attached thereto, along an electrical circuit to the cathode, and two electrons, returned to the cathode from the two hydrogen cations make it negative pole and the anode losing electrons becomes the positive pole, generating a direct electric current from the combined silicon semiconductor anode and metal grid to the cathode (Figure 27)

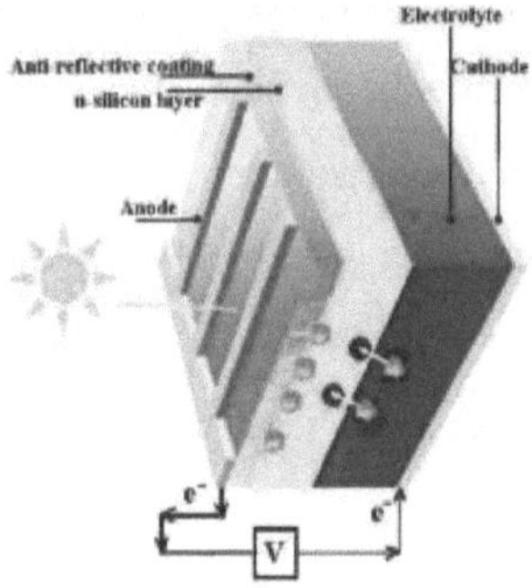

Figure 27: The combined N-type silicon semiconductor anode with attached metal grid operates under sunlight.

As established earlier [2], the size and intensity of the formation of negatively charged electrolytic hydrogen bubbles and positively charged oxygen bubbles can be controlled by increasing the electrical potential of the electrodes, making the intensely formed bubbles as small as required by the process.

The conclusion stated above was the basis for the development of a new photoelectrolyser-generator.

An important feature of the developed photoelectrolyser-generator, which produces pure hydrogen and oxygen at a below-market price, is a specially designed electrolysis base, which includes a membrane of fire-retardant material located between a combined silicon semiconductor anode with an attached metal grid, and an annealed graphite cathode, and a mechanism for adjusting the electrode gap at the bottom of the photoelectrolyser-generator.

A thin fire hose membrane placed between the combined silicon semiconductor anode with attached metal mesh and the cathode in the electrolysis base prevents the penetration of electrolytic oxygen bubbles formed by the attraction of four hydroxyl anions of the surrounding dissociated water giving up four electrons into four positively charged "holes" formed by the exposure of four photons of light to the combined silicon semiconductor anode with attached metal mesh

As a result of electrolysis in the cathode zone of the photoelectrolyser-generator, electrolytic hydrogen bubbles are intensively formed, which are removed from the cathode zone of the photoclectrolyser through a hydrogen tube in the form of a foam of gaseous hydrogen.

The arrangement of the electrolysis base, i.e. electrodes with a thin membrane and a mechanism for adjusting the gap between them, in the lower part of the photoelectrolyser-generator leads, in contrast to the photoelectrochemical cells existing today, where the electrodes are arranged vertically, to an intensive release of hydrogen and oxygen bubbles without the use of expensive electrolytes.

The arrangement of electrodcs in vertical direction, as it takes place in photoelectrochemical cells with aqueous electrolyte solution, in contrast to the electrolysis base located in the bottom of the photoelectrolyser-generator, where, as in all electroflotation devices, the electrodes are arranged in horizontal direction, occupying the whole bottom of the electroflotation apparatus, sharply reduces the cross-sectional area and increases the length of the conductive layer between the electrodes. The efficiency of a photoelectrochemical cell or a photoelectrolyser-generator is determined by the intensity of the outgassing in them.

According to Faraday's law, the mass of gas (***M***) released at the electrode during time (***t***) is equal to:

$$M = jigSt \qquad (2.4)$$

where ***j*** *is the* electrochemical equivalent of gas (for hydrogen j= 1.0-10-7 kg/(A^s); for oxygen j= 8.29-10-8 kg/(A^s)); ***i*** *is the* current density, A/m^2; ***g***=

90%, 95% is the current yield (ratio of actual gas mass transfer to theoretical calculated one in per cent); ***S*** is the cross-sectional area of the conductive layer between the electrodes, m^2.

A sharp decrease in the cross-sectional area and a sharp increase in the length of the conductive layer between the electrodes in accordance with (2.4) leads to a sharp decrease in the outgassing of hydrogen and oxygen in the photoelectrochemical cell or photoelectrolyser-generator and, as a consequence, to a sharp decrease in the efficiency of the photoelectrochemical cell or photoelectrolyser-generator.

To compensate for the reduction of hydrogen and oxygen outgassing in the aqueous electrolyte solution, expensive electrolyte (e.g., ***NaOH***) is used, the concentration of which in the solution is 40%, which leads to a sharp increase in energy consumption of the photoelectrochemical cell, reaching 4.5 kW^h/m3 to obtain hydrogen and oxygen.

A significant advantage of the developed photoelectrolyser-generator over existing electroflotation devices is the absence of electrodes in the electrolysis base, as which are used dense grids with fine cells, causing high currents of the order of 100 amperes at low voltages (20-30 V).

The explanation of this significant advantage is as follows.

If we consider the ***i*** - process current density as ***I/S***, where ***I*** is the constant process current, then equation (1.3.4) takes the following form:

$$M = jIgt \qquad (2.5)$$

From equation (2.5) it can be seen that the mass of gas released at the electrode during time ***t*** is directly proportional to the magnitude of the direct current and does not depend on the area of the electrode.

However, according to Ohm's law, the magnitude of the DC current of the process is directly proportional to the electrode voltage (***U***) and inversely proportional to the resistance of the interelectrode medium (***R***):

$$I = U/R \qquad (2.6)$$

The resistance of the interelectrode medium (***R***), in turn, depends on the interelectrode gap (***L***) and the cross-sectional area of the conductive layer between the electrodes (***S***) as follows:

$$R = \rho \frac{L}{S} \qquad (2.7)$$

Where ***ρ*** *is the* resistivity of the inter-electrode medium.

Substituting (2.7) into (2.6), we obtain:

$$I = (US/\rho L) \qquad (2.8)$$

Thus, it can be seen from (2.8) that a decrease in the cross-sectional area of the conductive layer between the electrodes of a photoelectrochemical cell or photoelectrolyser-generator (***S***), in our case a change in the shape of the electrode (silicon semiconductor with attached mesh anode), can be compensated by increasing the voltage (***U***) at the electrodes to maintain the same efficiency of the photoelectrochemical cell or photoelectrolyser-generator (the same gas content of the system - ***M***).

Increasing the voltage at the electrodes not only does not reduce the energy consumption efficiency of the photoelectrochemical cell, reaching 4.5 kW^h/m3 (for sunlight using photoelectrolyser-generator, the energy consumption efficiency reaches 0) to produce hydrogen and oxygen.

A significant advantage of the developed photoelectrolyser-generator over existing electroflotation devices is the absence of electrodes in the electrolysis base, as which dense meshes with fine cells are used, causing large currents of the order of 100 amperes at low voltages (20-30 V).

The arrangement of electrodes in the vertical direction not only sharply reduces the cross-sectional area and the length of the conductive layer between the electrodes, which leads to a sharp decrease in the outgassing and intensity of the formation of electrolytic hydrogen and oxygen bubbles, but also prevents the formation and rapid removal of electrolytic hydrogen and oxygen bubbles through the hydrogen and oxygen tubes, respectively.

Electrolytic bubbles of hydrogen or oxygen formed on the entire horizontal surface of the electrode or anode, breaking off from the electrode surface, rush upwards and in the form of gaseous hydrogen bubbles or oxygen bubbles are removed from the photoelectrolyser-generator through hydrogen or oxygen tubes under the action of Archimedes' force acting in a strictly vertical direction in relation to the cathode or anode.

In vertically located electrodes, the resulting bubbles of hydrogen and oxygen floatup near the cathode and anode, forming a narrow flow due to the Archimedes force acting tangentially to the cathode or anode, preventing the separation of hydrogenbubbles from the cathode or oxygen bubbles from the anode.

In the developed photoelectrolyser-generator is carried out round-the-clock continuous production of hydrogen and oxygen for uninterrupted movement of the vehicle, using the same technology as described above, i.e. using in this case as lighting photoelectrolyser-generator solar light from the light receiver, transmitted via optical fibre, and as a useful electrical load of photoelectrolyser-generator - LED (3) with a battery, charged in the process of lighting photoelectrolyser-generator.

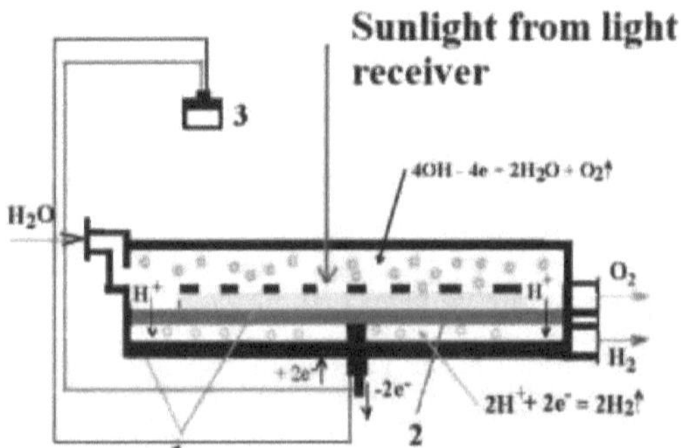

Figure 28: 24/7, continuous hydrogen and oxygen production for an underground life support system

In this case, the photoelectrolyser-generator electrical payload is an LED with a battery charged during the process of illuminating the photoelectrolyser-generator with sunlight during the day, which starts illuminating the photoelectrolyser-generator in the absence of sunlight at night until the battery is completely discharged in the morning.

In this case, the circuit diagram of the LED control board will look as shown in Figure 29.

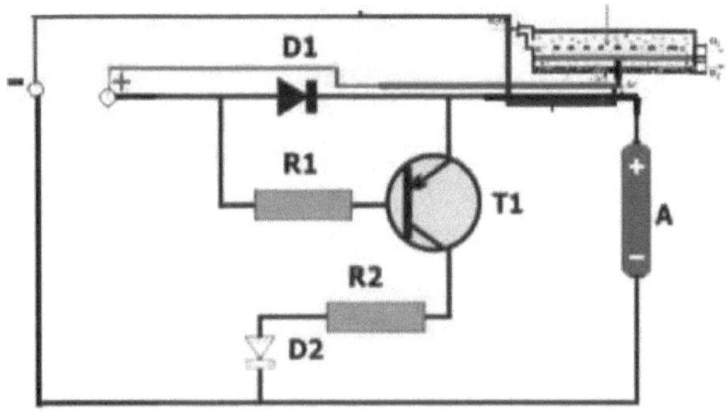

Figure 29: Circuit diagram of the LED control board.

The current produced by the photoelectrolyser-generator charges the battery (A) via diode D1.

The positive potential applied to the base via resistor R1 "keeps" transistor T1 switched off and LED D2 does not light up.

When the illumination of the anode of the photoelectrolyser-generator anode is significantly reduced, the transistor opens (due to the reduction of the positive potential applied to the base) and connects the LED D2 to the battery.

The LED starts to glow, illuminating the anode of the photoelectrolyser-generator.

Diode D1 prevents the battery from discharging during light illumination.

When the light illumination is switched on, the positive voltage coming from the "+" terminal of the photoelectrolyser-generator anode to the base "closes" the transistor T1 and the LED D2 stops glowing and the battery starts charging.

Continuous charging of the vehicle motor fuel cell for round-the-clock uninterrupted movement of the vehicle is carried out as follows.

The resulting gases are separately fed: hydrogen to the porous hydrogen electrode (anode) and oxygen to the porous oxygen electrode (cathode) of the fuel cell (Figure 30).

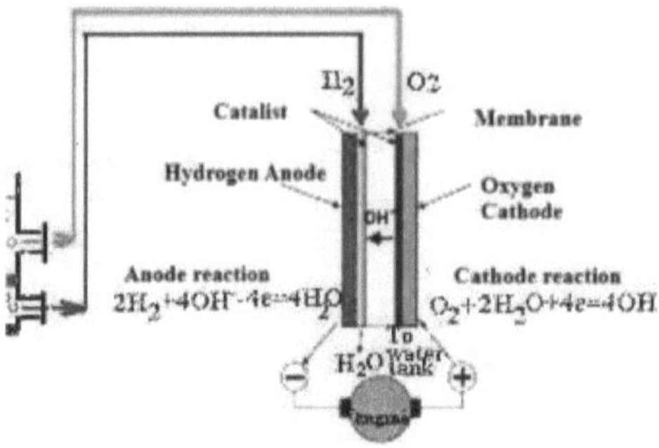

Figure 30: The resulting gases separately flow hydrogen into the porous hydrogen electrode (anode), oxygen into the porous oxygen electrode (cathode) of the fuel cell.

The anode and cathode are separated by a membrane.

Each of the electrodes is coated with a layer of catalyst.

As a result, four hydrogen molecules in the hydrogen electrode lose four electrons, creating a negative potential at one of the motor contacts, and reacting with four ***OH⁻*** anions derived from the oxygen electrode by reacting an oxygen molecule with two water molecules, releasing four positively charged particles, creating a positive potential at the other contact of the vehicle motor, forming four water molecules (water vapour):

$$O_2 + 2H_2O + 4e \rightarrow 4OH- \rightarrow 2H_2 + 4OH^- - 4e \rightarrow 4H_2O \qquad (2.9)$$

As a result, the vehicle's electric motor is started and the generated water vapour flows back to the conventional water tank, closing the cycle.

Thus, the developed 24/7 hydrogen generator produces pure hydrogen and oxygen, which continuously charges the (energy efficient) fuel cell of the vehicle motor.

Figure 31 shows a complete cycle of 24/7 hydrogen and oxygen production and charging of a vehicle motor fuel cell using the previously developed photoelectrolyser-generator, which uses sunlight (3) from a light receiver transmitted via optical fibre as illumination of the photoelectrolyser-generator, and an LED (4) with a battery charged while illuminating the photoelectrolyser-generator with sunlight during the day, which starts illuminating the photoelectrolyser-generator

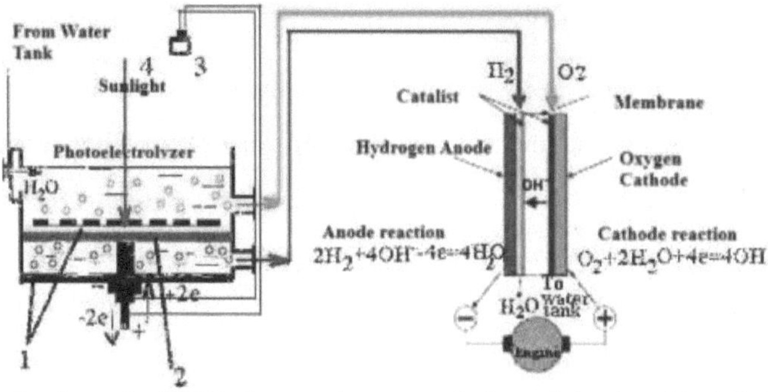

Figure 31: Complete cycle of 24/7 hydrogen and oxygen production and charging of the fuel cell of the life support system's electric generator electrolyser-generator

Thus, the developed round-the-clock and continuous production of hydrogen and oxygen in the photoelectrolyser-generator allowed the author to carry out round-the-clock and continuous charging of the fuel cell of the electric motor of the vehicle for uninterrupted movement of the vehicle, provided by the use of sunlight from a light receiver transmitted via optical fibre as illumination for the photoelectrolyser-generator in , and as a useful electrical load of the photoelectrolyser-generator - light from a light receiver transmitted via optical fibre.

Figure 32 shows an electric car with a natural light system using sunlight travelling through an optical fibre installed on its roof.

Figure 32: An electric car with a natural lighting system using sunlight passing through an optical fibre installed on its roof.

CONCLUSIONS.

1. The use of a light receiver that dynamically tracks the light activity of sunlight collected during the day, transmitted via optical fibre to the bomb shelter to illuminate the rooms, which charges the batteries of photocells located in the most illuminated areas of these rooms, the electric current from which at night turns on the LEDs located next to the interior lights of the rooms, until the batteries are completely discharged in the morning, makes it possible to provide round-the-clock illumination of the rooms of the bomb shelter.

2. Twenty-four hour, continuous production of hydrogen and oxygen for the bomb shelter life support system in a specially designed photoelectrolyser-generator is achieved by using as light for the photoelectrolyser-generator sunlight from a light receiver transmitted via optical fibre, and an LED with a battery charged during the day by illuminating the photoelectrolyser-generator with sunlight, which starts illuminating the photoelectrolyser-generator in the absence of sunlight at night until the batteries are fully discharged

3. Twenty-four-hour production of oxygen and hydrogen in a specially designed photoelectrolyser-generator enables 24-hour, continuous charging of the fuel cell of the life support system's electrical generator for uninterrupted power supply to bomb shelters.

4. Round-the-clock, continuous production of hydrogen and oxygen in a specially designed photoelectrolyser-generator, achieved by using as illumination of photoelectrolyser-generator sunlight from a light-receiver transmitted via optical fibre, and LED with a battery, charged during the day by illumination of photoelectrolyser-generator by sunlight, which begins to illuminate photoelectrolyser-generator in the absence of sunlight at night until the battery is fully discharged in the morning, and not

LINKS

1. Get healthy office lighting with natural light using products from Parans.https://parans.com/product/

2. Shoikhedbrod M. 24/7 water electrolysis process under the influence of light energy, Lambert, 2024.

3. Goff D. Fiber optic reference guide, Routledge, 2002. https://www.taylorfrancis.com/books/mono/10.4324/9780080506319/fiber-optic-reference-guide-david- goff

4. Crispto J. Introduction to fibre optics, Elsevier, 2005. https://books.google.com/books/about/Introduction_to_Fiber_Optics.html?hl= ru &id=RP2FQw5uSywC#v=onepage&q&f=false

5. Udd E. Fiber optic smart structures, Fiber Optic Scientists, Sensors: An Introduction for Engineers and 2011; 373-398. https://onlinelibrary.wiley.com/doi/abs/10.1002/9781118014103.ch14

6. DeCusatis C., DeCusatis J.S. Fiber optic essentials, Elsevier, 2010. https://books.google.com/books/about/Fiber_Optic_Essentials.html?hl= ru&id=c-B990MwuPkC#v=onepage&q&f=false

7. Hecht J. City of light: the story of fibre optics, Oxford University Press, USA, 2004. https://books.google.com/books/about/City_of_Light.html?hl=ru&id=4oMu7RbGpqUC#v= onepage&q&f&f=false

8. Campagnola L., Wang H., Zylka M. Fiber-coupled light-emitting diode for localised photostimulation of neurons expressing channelrhodopsin-2, Journal of neuroscience methods 2008; 169 (1): 27-33. https://www.sciencedirect.com/science/article/abs/pii/S0165027007005651

9. Kapany N. Fiber optics, Scientific American, 1960; 203 (5): 72-81. https://www.jstor.org/stable/24940695

10. Nelson J. The physics of solar cells, *World Scientific Publishing Company*,2003.

11. Bagher A., Vahid M., Mohsen M. Types of solar cells and

applications,American Journal of optics and photonics, 2015; 3 (5):94-113.

12. Fraas L., Partain L. Solar cells and their applications, *John Wiley & Sons,*2010.

13. Strandwitz N.C., Comstock D.J, Grimm R.G., Nielander A.C., Elam J.,Lewis N.S.Photoelectrochemical behaviour of n-type Si (100) electrodescoated with thin films of manganese oxide grown by atomic layerdeposition, *The Journal of Physical Chemistry*, 2013;C 117(10): 4931-4936.

14. Nielander A.C., Bierman M.J., Petrone N., Strandwitz N.C., Ardo S., Yang F., Hone J., Lewis N.S. Photoelectrochemical behaviour of n-type Si (111) electrodes coated with a single layer of grapheme, Journal of the AmericanChemicalSociety, 2013; 135(46): 17246-17249.

FSC
www.fsc.org
MIX
Papier | Fördert gute Waldnutzung
FSC® C083411

Kolibri 360 GmbH
Ferdinand-Jühlke-Straße 7
99095 Erfurt, Deutschland
produktsicherheit@kolibri360.de

Druck:
CPI Druckdienstleistungen GmbH
im Auftrag der
Kolibri 360 GmbH
Ferdinand-Jühlke-Str. 7
99095 Erfurt